AF897177

EARTH'S LANDFORMS

CANYONS

by Lisa J. Amstutz

PEBBLE
a capstone imprint

Pebble Emerge is published by Pebble, an imprint of Capstone.
1710 Roe Crest Drive
North Mankato, Minnesota 56003
www.capstonepub.com

Copyright © 2021 by Capstone. All rights reserved. No part of this publication may be reproduced in whole or in part, or stored in a retrieval system, or transmitted in any form or by any means, electronic, mechanical, photocopying, recording, or otherwise, without written permission of the publisher.

Library of Congress Cataloging-in-Publication Data is available on the Library of Congress website.
ISBN: 978-1-9771-2458-6 (library binding)
ISBN: 978-1-9771-2634-4 (paperback)
ISBN: 978-1-9771-2501-9 (eBook PDF)

Summary: Describes canyons, including their features, how they are made, and where they can be found.

Image Credits
Capstone Press/Karon Dubke, 20; Dreamstime/Chong Chen, 10; Shutterstock: Art Boardman, 18, Charles Knowles, 4, Eugene Fedorenko, 17, Evgeny Haritonov, 15, Fernando Tatay 11, Jason Patrick Ross, cover, Jeroen Mikkers, 13, Kirill Skorobogatko, 14, Kris Wiktor, 16, Lukas Bischoff Photograph, 5, Maridav, 19, Maverick, 9, MNStudio, back cover, ShuPhotography, 8, TK iStock, 7

Editorial Credits
Editor: Alesha Sullivan; Designer: Bobbie Nuytten, Media Researcher: Kelly Garvin; Production Specialist: Tori Abraham

All internet sites appearing in back matter were available and accurate when this book was sent to press.

Printed and bound in China.
003322

Table of Contents

Words in **bold** are in the glossary.

WHAT IS A CANYON?

Have you ever seen a canyon? It is a deep area with **steep** sides. Rocky walls are on each side. Sometimes a river is at the bottom of a canyon.

A canyon is one kind of **landform**. There are many types of canyons. Some are small and short. Others are big and long.

HOW ARE CANYONS MADE?

Many canyons are made by rivers. Water in a river moves to a lake or the ocean. Rushing water breaks down rocks and land. They slowly wear away.

Ice and rain can also make canyons. Water freezes inside rock cracks. This pushes the rocks apart. When it rains, the rocks break down. The canyon gets wider.

Bursts of rushing water can make slot canyons. These canyons are usually not very wide.

A box canyon starts at the bottom of a **cliff**. Over time, water wears away the rock. The cliff crumbles and falls. This type of canyon has three sides.

WHERE ARE CANYONS FOUND?

Canyons are all over the world. Some are in the ocean. Some are on mountaintops. The world's deepest canyon cuts through mountains in China. It is more than 3.3 miles (5.3 kilometers) deep.

The Grand Canyon is a famous canyon in the United States. It is about 1 mile (1.6 km) deep. It is 277 miles (446 km) long.

ALL ABOUT CANYONS

Some canyons have taken millions of years to form. Layers of rock build up over time on the sides of a canyon. Look closely! These layers can be many colors. Different types of **minerals** in the rock give canyons their color.

People study canyons to learn about the past. The walls show how the canyon was made. Sometimes people find **fossils** in canyon rock layers. These remains show what animals and plants lived there millions of years ago.

fossil in canyon

LIFE IN A CANYON

Plants and animals live in and by canyons. It can be hard for plants to grow on the steep, rocky walls. Some grasses and bushes can grow at the top or bottom. Bighorn sheep climb the rocks. Some birds make nests in canyon walls.

bighorn sheep

Some people live close to canyons. Others take trips to visit them. They hike trails. They take boats on rivers through canyons. Would you like to see the colorful rocks in a canyon?

Make a Canyon

With just a few materials, you can make a canyon of your own!

You'll Need:

- colored paper
- white glue
- colored markers

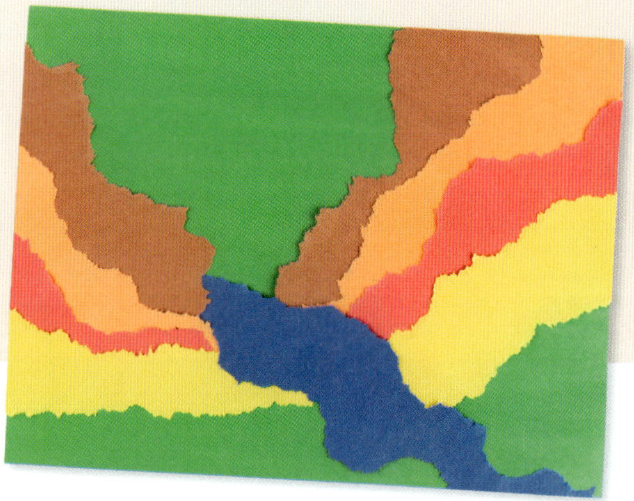

Instructions:

1. Use a piece of paper for your background.

2. Tear pieces of paper into narrow strips. Use different colors of paper, such as red, orange, light brown, and gray.

3. Put them on the paper so they make a V shape like a canyon. Glue the pieces of paper down. If you'd like, tear a blue strip of paper. Glue it at the bottom of your canyon to make a river.

4. You can add more details with markers, such as a sky and plants.

Glossary

cliff (KLIF)—a high, steep rock face

fossil (FOSS-uhl)—the remains or traces of an animal or plant that lived millions of years ago

landform (LAND-form)—a natural feature of the land

mineral (MIN-ur-uhl)—a material found in nature that is not an animal or a plant

steep (STEEP)—having a sharp slope or slant

Read More

Kalman, Bobbie. *What Are Landforms?* New York: Crabtree Publishing Company, 2018.

Terp, Gail. *Amazing Canyons Around the World.* North Mankato, MN: Capstone, 2019.

Internet Sites

National Geographic Kids: Grand Canyon National Park
https://kids.nationalgeographic.com/explore/nature/grand-canyon/

PBS LearningMedia: Landforms
https://www.pbslearningmedia.org/resource/buac17-k2-sci-ess-landforms/landforms/

Index